UNDERSTANDING
CORONAVIRUS (COVID-19)

A REVISED 2021 EDITION

TONNY RUTAKIRWA

Published by Tonniez Publishing Press,
a member of Tonniez Group of Companies

c/o Tonniez Group Holdings,
86-90 Paul Street, London, EC2A 4NE

Email: books@rutakirwa.com
Websites: https://www.TonnyRutakirwa.com
https://www.Rutakirwa.com
https://www.TonniezGroup.com

ISBN (eBook): 978-1-008-99555-0
ISBN (Paperback): 978-1-008-99554-3
ISBN (Paperback): 978-1-80049-478-7
ISBN (Hardback): 978-1-008-99553-6
ISBN (Hardback): 978-1-80049-479-4

Editor: Tekky Andrew-Jaja
Layout and Design: Rica Cabrex

Contents

CHAPTER 1

Introducing You to The Renowned Coronavirus

▨ WHAT IS CORONAVIRUS?

Coronavirus disease is a virus that is part of a large family of viruses called coronaviruses (CoV). It has been known to cause common respiratory infections ranging from common diseases like the common cold to more critical illnesses such as severe acute respiratory syndrome (SARS) and Middle East Respiratory Syndrome (MERS). However, the Coronavirus disease (COVID-19) is a new strain of the coronaviruses that was discovered in 2019 that has never been previously identified in humans.

Coronaviruses are known to be zoonotic—that is they are usually transmitted between humans and animals.

Several investigations proved that SARS was transmitted to humans from civet cats, and MERS was also transmitted from dromedary camels to humans. There are still several strains of coronaviruses that are known in animals that are yet to infect humans.

Coronaviruses got their name from the resemblance that the spikes protruding from their surface have with a crown and the sun's corona. They infect humans and animals, thereby causing illness in the respiratory tracts. Every year, there are at least four different strains of coronaviruses that cause mild infections such as the common cold. Most people will get infected with at least one of these viruses at a point in their lives.

There was a strain of the coronaviruses that circulated in China back in 2003—this strain of coronavirus was called SARS. The virus was contained after it killed about 774 and made over 8,000 people sick. Following that, in 2012, another strain of the coronaviruses called MERS plagued Saudi Arabia. However, the coronavirus disease (COVID-19) currently plaguing the world originated in Wuhan, China.

COVID-19 is usually rutransmitted via respiratory droplets from sneezing and coughing. Also, it is considered to be the most contagious when people have become symptomatic. However, there is a possibility of transmission even before symptoms begin to appear.

How Dangerous is the Coronavirus Disease?

Accurately assessing the lethality of a new virus is not an easy task. COVID-19, however, appears to be less fatal than SARS or MERS but more fatal than the seasonal flu. Studies have shown that the fatality rate of COVID-19 is about 2%. However, statistics show that about 5% of the patients infected with the COVID-19 had critical illnesses.

Although children are less likely to be infected with the COVID-19, middle-aged and other adults have more chances of being infected. Also, compared to women, men are more likely to die from COVID-19. This is due to men producing weaker immune responses and having higher tobacco consumption rates. Additionally, having type 2 diabetes and high blood pressure increases the risk of having complications when infected by the COVID-19.

THE HISTORY OF CORONAVIRUSES

The history of human coronaviruses is said to have begun in 1965 as Tyrrel and Bynoe discovered that they were able to passage a virus called B814. This virus was found in the embryonic tracheal organ cultures that were obtained from the respiratory tract of a human adult infected with the common cold. They were able to demonstrate the presence of an infectious agent. This was done by inoculating the medium from these cultures in human volunteers intranasally.

Although the common cold was produced in a significant number of the subjects, Tyrell and Bynoe were not able to grow the agent in tissue cultures. However, Hamre and Procknow were able to grow a virus. They named 229E in tissue culture with samples they got from medical students with colds. Both viruses were found to ether-sensitive, and they required a lipid-containing coat for infectivity. Similarly, the viruses were found not to be related to any known myxoviruses or paramyxoviruses. These viruses were termed "OC" to show they grown in organ cultures.

However, Almeida and Tyrrell performed electron microscopy on the fluid from the organ culture they infected with the virus B814, and they discovered particles that looked like the bronchitis virus that infects chickens. The OC viruses were all found to have the same morphology.

Tyrrell was leading a group of virologists in the late 1960s, that was working with several animal viruses and the human strain. The infectious bronchitis virus, gastroenteritis virus of swine, and mouse hepatitis were all viruses that were included in the study. However, the study discovered that the morphology of these viruses was the same as those seen in the electron microscopy done by Almeida and Tyrrell.

Therefore, this new group of viruses was named coronaviruses due to the crown-like appearance of their surface projections, and they were later accepted officially as a new genus of viruses. Research using serologic techniques has given a considerable amount of information as regards coronaviruses that affect the human respiratory system. Also, it was discovered that coronavirus infections occur more often in the spring and winter than in the fall and summer due to temperature.

Over the years, the human strains OC43 and 229E were the strain of coronaviruses that were studied. However, along the line, it was discovered there are other strains of coronaviruses. For example, Bradburne discovered a coronavirus strain called B814. Also, this was seen to not been serologically identical to either OC43 or 229E.

Studies found that coronaviruses were associated with different respiratory illnesses, even though their pathogenicity is low. However, the predominant illnesses that were associated with coronaviruses were upper respiratory infection as well as the occasional cases of pneumonia in children and young adults. Coronaviruses were also known to produce asthma exacerbations in children, plus chronic bronchitis in adults and older adults.

EPIDEMIOLOGY

On the 31st of December, 2019, a cluster of pneumonia cases having unknown causes was reported to the authorities in Wuhan, China. As in early January 2020, an investigation was launched to determine the causes of these cases. These cases had been linked to the Huana seafood wholesale market that sold live animals, which made them think the virus might be zoonotic.

The virus that caused the outbreak was discovered to be SARS-CoV-2. This happened to be a new virus that is closely related to coronaviruses found in bat, SARS-CoV, and pangolin coronaviruses. The new strain of coronaviruses was believed to have originated in horseshoe bats.

The first person with symptoms related to this virus was traced as far back as the 1st of December, 2019. However, the person had no connections with the later cluster linked to the market. Among the earlier reported cluster of cases reported in December 2019, about 2/3rds of the cases were found to have links to the market. However, on the 14th of March, 2020, South China's morning post gave an unverified report about a 55-year-old man from Hubei province being the first person to contract the virus as far back as the 17th of November, 2019.

On December 24th, 2019, a bronchoalveolar lavage fluid (BAL) was sent by Wuhan Central Hospital to Vision Medicals for further analysis, since Wuhan Central Hospital had been unable to resolve the case. Since Vision Medicals is a sequencing company, it was expected to unravel the functional and structural elements of the associated disease-causing organism in the sample. Three days after the sample was sent, Vision Medicals gave their results to Wuhan Central Hospital and further notified the Chinese CDC that the sample gave indications of a new coronavirus.

On December 30th, the Wuhan Municipal Health Commission gave a public address concerning a pneumonia outbreak. By then, 27 cases had been confirmed, and this triggered an investigation. On December 31st, the World Health Organization learned about the pneumonia cases in Wuhan, and launched an investigation in January 2020.

A fact-finding mission by a joint task force between the World Health Organisation (WHO) and China revealed that the epidemic peaked in China between late January and the beginning of February 2020 as the virus spread to other provinces especially due to the Chinese New Year migration. After the first case was reported in Wuhan, at the end of 2019, there had been over 110 million reported cases of COVID-19 worldwide.

By the 26th of February, 2020, WHO gave a report that showed new cases dropping in China. However, it seemed to be increasing in Italy, South Korea, and Iran. This made the number of new cases without China exceed the number of new cases within China for the first time since the beginning of the outbreak. These cases were initially occurring among travelers from China or people who were in contact with travelers from China.

The reported numbers may substantially have been lower than the actual number of cases, especially among those with very mild symptoms. Also, the reported number reflected the local decision on when and whom to test. For instance, the UK reported 798 confirmed cases on March 13th, 2020. However, health officials estimated the actual number of people infected by the virus to be somewhere between 5,000 and 10,000.

By mid-February 2021, things had taken a new change and the numbers had changed drastically, with China moving out of the top 80 countries in terms of total cases. The leading country was the United States with more than 28 million total cases followed by India at almost 11 million cases and Brazil coming third with almost 10 million cases.

Fatality Rate of The Virus

The fatality rate of COVID-19 has been estimated to be about 1.4%. However, the rate is reported to vary from country to country. By studying those who had died from the virus, it was found out that the time from the development of symptoms to the time of death is usually between 6 and 41 days. As of February 17th, 2021, there have been over 2,434,000 recorded deaths from the infection of COVID-19.

The majority of individuals who have died from the virus are the elderly. About 80% of deaths were in those that were above 60. However, 75% of the deaths were from those that had pre-existing health conditions, which include diabetes and cardiovascular disease.

The first recorded death from COVID-19 was on the 9th of January, 2020, in Wuhan, while the first death to be recorded outside China was on the 1st of February, 2020, in the Philippines. The first death to be recorded outside China was in France. By March 13th of the year 2020, more than 40 countries on every continent except Antarctica had reported deaths.

THE NAMING OF COVID-19

COVID-19 has been referred to by many names, most of them unofficial. During the early days of its outbreak, the virus was commonly referred to as "coronavirus." However, this is a broad term since it encompasses a group of RNA viruses that cause respiratory tract infections such as common cold, SARS, MERS, and COVID-19. At the same time, the virus was also being referred to as "Wuhan coronavirus." This name was created out of the first geographical location in which the virus appeared. The first case of COVID-19 was identified in Wuhan, and this led to the "Wuhan coronavirus" name. This naming based on geographical locations is not a new thing, as it was also used in naming the Zika virus, Spanish flu, and

Middle East Respiratory Syndrome. Other names that were oriented towards the geographical location include "China coronavirus" and "Wuhan pneumonia."

The first official name to be recommended for the virus was 2019-nCoV while the first recommended name for the disease was 2019-nCoV acute respiratory disease. The names recommended by the World Health Organization are in strict adherence to the 2015 guidance and international guidelines against using various factors such as geographical locations in the naming. These guidelines are aimed at preventing social stigma that may emanate when diseases are named using geographical locations or other aspects that put a group of people or animals on the receiving end.

The official name for the virus (COVID-19) was issued by the World Health Organization on 11th February 2020. The was created as follows:

CO: Corona
VI: Virus
D: Disease
19: 2019, when the disease was first identified.

CHAPTER 2

Causes, Diagnosis, and Prevention of Coronavirus

CAUSES OF THE CORONAVIRUS DISEASE

It is still unclear how contagious COVID-19 is. However, there have been some speculations about how it is transmitted as well as its virology

TRANSMISSION OF THE DISEASE

At the beginning of the investigation into the virus, it was discovered that most patients had either worked in or visited the seafood market that sold live animals. Subsequently, as the outbreak progressed, the primary mode of transmission of the virus then became person-to-person spread. The person-to-person spread of the virus is said to occur mainly through respiratory droplets, just like the case of influenza.

The virus is released in the respiratory secretions whenever an infected person sneezes, coughs, or talks—it can infect another person if it makes contact with the mucous membranes. Also, infections can occur if a person touches his or her mouth, eyes, or nose after touching an infected surface. The droplets where the virus resides do not travel for more than six feet and cannot linger in the air.

Given the uncertainty with the transmission mechanism of the virus, taking airborne precautions is recommended routinely. According to several studies, it was discovered that the rate of transmission from individuals with symptomatic infections varies by infection control interventions and location. In China, the transmission rate was 1–5% among tens of thousands who had close contacts with an infected person while in the USA, the transmission rate was 0.45% among 445 people that had close contacts with infected persons.

Transmission of COVID-19 can also occur from asymptomatic individuals (not displaying symptoms) or individuals within the incubation period. However, the extent and rate at which this happens remain yet to be known. Although the virus can be detected in blood and stool specimens, according to the WHO-China report, transmission through fecal-oral does not appear to be a significant factor in the spread of the virus.

Virology

After carrying out a phylogenic analysis and full-genome sequencing, it was found out that the coronavirus that causes COVID-19 is a beta coronavirus. This belongs to the same subgenus as several bat coronaviruses as well as a severe acute respiratory syndrome, but it belongs to a different clade. However, the structure of the receptor-binding gene region of the COVID-19 is similar to the SARS coronavirus. Also, the virus has been shown to uses the same receptor as well as using the angiotensin-converting enzyme 2 (ACE2) for cell entry.

Therefore, the Coronavirus Study Group of the International Committee on Taxonomy of Viruses proposed that COVID-19 be called "severe acute respiratory syndrome coronavirus 2" (SARS-CoV-2). The closest RNA sequence that was similar to that of the COVID-19 is from two bat coronaviruses—bats are apparently the primary source of COVID-19. However, it still remains unknown if the virus was transmitted directly from bats or through an intermediate host or some other mechanism.

Further analysis of the strains of COVID-19 from China showed that there are two different types of COVID-19. These were designated type L and type S. The type L is accounting for 70% of the strains predominated in the early days of the epidemic in China. However, it is accounting for a lower proportion of the strains outside Wuhan. The type S that accounted for 30% of the strain in Wuhan has begun to account for more outside Wuhan. The clinical implication of these findings are, however, uncertain.

DIAGNOSIS OF THE CORONAVIRUS DISEASE

After being exposed to the virus or beginning to develop symptoms of COVID-19, you are advised to contact your doctor as soon as possible. Your doctor will then determine whether to conduct the test for COVID-19 based on the signs and symptoms you are exhibiting. Having had close contact with someone who has been diagnosed with COVID-19 or having been to any area where there is an ongoing community spread of COVID-

19 in the last 14 days are also determinants to know whether the test for COVID-19 is necessary.

However, the confirmation of COVID-19 is through reverse transcription-polymerase chain reaction (rRT-PCR) of infected secretions, which has about 71% sensitivity, and CT imaging, which has about 98% sensitivity.

Viral Testing

On the 17th of January, WHO published different RNA testing protocols for COVID-19. The testing for this virus is said to use a real-time reverse transcription-polymerase chain reaction (rRT-PCR). The tests are done on respiratory or blood samples, with the results being ready from within a few hours to a few days.

CT Imaging

The CT imaging features on computed tomography and radiographs have been described in a limited number of cases. However, the Italian Radiological society is compiling the imaging findings for confirmed cases in an international online database. Unfortunately, due to the overlap this imaging has with other infections such as adenovirus, CT imagining without confirmation by rRT-PCR is of little or no use in identifying COVID-19.

Signs and Symptoms of the Coronavirus Disease

The symptoms of COVID-19 are not specific, and the individuals who are infected may be asymptomatic, or they could develop flu-like symptoms like fever, fatigue, shortness of breath, cough, or muscle pain. However, pneumonia is the most severe frequent manifestation of an individual infected with COVID-19—it is usually characterized by bilateral infiltrates, cough, fever, and dyspnea. Unfortunately, there has been no clinical feature that can be used to differentiate COVID-19 from other viral respiratory infections.

The most common symptoms at the onset of the illness according to the study are as follows:

- Fever in 87.9%
- Dry cough in 67.7%
- Fatigue in 38.1%
- Sputum production in 33.4%
- Shortness of breath in 18.6%
- Muscle pain or joint pain in 14.8%
- Sore throat in 13.9%

- Headache in 13.6%
- Chills in 11.4%
- Nausea or vomiting in 5.0%
- Nasal congestion in 4.8%
- Diarrhea in 3.7%
- Haemoptysis in 0.9%
- Conjunctival congestion in 0.8%

The further development of the disease could lead to acute respiratory distress syndrome, septic shock, sepsis, and finally, death.

Incubation Period

The incubation period for COVID-19 is believed to be within 14 days after exposure. However, most cases were found to occur after 4–5 days following exposure. By using data from 181 cases in China, it was discovered that symptoms would develop in about 2.5% of the infected individual after 2.2 days, and the remaining 97.5% of infected individuals develop symptoms after 11.5 days. This brings the median incubation period to about 5.1 days.

Severity of Illness

The severity of infection of COVID-19 ranges between mild and critical—most cases were found not to be severe. A report from the Chinese Center for Disease Control and Prevention showed the statistics of the severity of cases:

- Mild: About 81% of the cases were reported to be mild, which is having no or mild pneumonia.
- Severe: About 14% of the cases were reported to be severe, which is having dyspnea or hypoxia.
- Critical: About 5% of the cases were reported to be critical, which is having respiratory failure, shock, or multiorgan dysfunction.

The fatality rate for the cases was about 2.3%, as no deaths were reported among non-critical cases.

PREVENTION OF THE CORONAVIRUS DISEASE

Coming up with effective ways to prevent the spread of COVID-19 is a global challenge, as viruses are only preventable through antiviral vaccinations. However, coming up with and distributing safe and effective vaccines takes time, which means a vaccine for COVID-19 is going to take some time before it is fully distributed.

Already approved vaccines include Pfizer, Moderna, J&J and Oxford, AstraZeneca.

However, there are some ways to prevent COVID-19 from continually spreading. WHO and CDC, however, recommend several precautions to avoid being infected with COVID-19. There are as follows:

- Make sure you wash your hands for at least 20 seconds using soap and water.
- Use a hand sanitizer that contains at least 60% alcohol.
- On occasions where your hands are not clean, avoid touching your mouth, eyes, and nose.
- When you are sick, do not share glasses, dishes, bedding, and any household item.
- In events where you are sneezing or coughing, cover your mouth with your elbow or tissue. Make sure to throw away any and all used tissues.
- Avoidance of mass gathering and large events.
- Avoid close contact with individuals that are sick or symptomatic; make sure to stay at least six feet away from them.
- In an event where COVID-19 has begun to spread in your community, keep a safe distance between you and others, especially if you have a high risk of serious illness.

- If you are sick or showing symptoms, stay home from school, work, or any public area.
- Disinfect the surfaces you touch regularly.
- Wear a face mask to prevent any respiratory illness.
- Avoid eating undercooked or raw meat.
- Avoid contact with live animals or surfaces those animals might have touched.

Many governments have restricted and advised against non-essential travel to and from the countries or places that are affected by the pandemic. Also, health care practitioners taking care of persons who have been infected by the COVID-19 are to use standard precautions, airborne precaution, contact precaution as well as eye protection.

Contact tracing was a primary method that health authorities used to trace the source of infection to prevent further transmission during the early stages. Unfortunately, myths are circulating about how to prevent being infected by COVID-19 by raising the nose or gargling with mouthwash.

Social Distancing

Social distancing is a prevention action to reduce the spread of disease by minimizing close contact among individuals. The methods used to achieve this include travel restriction, quarantines, and closure of schools, stadiums, theatres, shopping centers, and workplaces. Individuals are advised to apply social distancing methods by staying at home, avoiding crowded areas, use of no-contact greeting, limiting travel, and physically distancing themselves from others.

Due to the pandemic, many governments are mandating and recommending social distancing in regions affected by the outbreak. The elderly and those with underlying medical conditions like respiratory disease, hypertension, heart diseases, and so on have been advised to stay at home as much as possible.

Self-Isolation

This is being mandated for individual who have been diagnosed with COVID-19 as well as those that are suspected of being infected. Health agencies have issued detailed instructions on how to properly self-isolate. Also, self-isolation has been mandated by several governments for the population living around the infected areas.

Anyone who might have been exposed to COVID-19 or who recently traveled to a country with a wide transmission rate is mandated to self-quarantine for 14 days to ensure they are not infected.

Surgical Mask

Masks are being recommended for use for those who are at high risk—for example, those taking care of individuals infected with COVID-19. Additionally, wearing a mask helps people to avoid touching their mouths or noses.

Surgical masks are designed mainly to protect people from the wearer of the mask. However, some masks protect the wearer, but they are called respirators.

CHAPTER 3

How Coronavirus Has Affected the World So Far

▨ STATISTICS CONCERNING THE CORONAVIRUS DISEASE

As of February 17th, 2021, there have been more than 110 million reported cases of COVID-19 worldwide. China, where the virus was first detected, currently stands at position 84 with 89,795 reported cases. 219 countries and territories have reported at least one COVID-19 case.

Since the virus began to spread from China all over the world, the WHO was prompted to designate the virus outbreak as a pandemic. There were human-to-human transmissions among individuals who had no apparent link to China.

Coronavirus Disease Cases

As of February 17th, 2021, there have been a total of 110,223,306 confirmed cases and 2,434,379 deaths from the COVID-19 outbreak. Following is a breakdown of the cases for the top 50 countries:

Country, Other	Total cases	New cases	Total death	New death	Total recovered	Active cases
World	110,223,306	+197,772	2,434,379	+5,923	85,094,677	22,694,250
USA	28,391,050	+9,830	500,190	+199	18,483,906	9,406,954
India	10,949,301	+12,195	156,033	+84	10,654,146	139,122
Brazil	9,921,981		240,983		8,883,191	797,807
Russia	4,112,151	+12,828	81,446	+467	3,642,582	388,123
UK	4,071,185	+12,718	118,933	+738	2,282,703	1,669,549
France	3,489,129		82,812		244,238	3,162,079
Spain	3,107,172	+10,829	66,316	+337	*2,411,183*	629,673
Italy	2,751,657	+12,074	94,540	+369	2,268,253	388,864
Turkey	2,602,034		27,652		2,489,624	84,758
Germany	2,356,829	+4,063	66,769	+233	2,154,600	135,460
Colombia	2,202,598		57,949		2,095,105	49,544
Argentina	2,033,060		50,432		1,838,291	144,337
Mexico	2,004,575	+8,683	175,986	+1,329	1,563,992	264,597
Poland	1,605,372	+8,694	41,308	+279	1,354,598	209,466
Iran	1,542,076	+8,042	59,184	+67	1,317,612	165,280
South Africa	1,494,119		48,313		1,396,951	48,855
Ukraine	1,280,904	+4,286	24,689	+147	1,128,890	127,325
Peru	1,244,729		44,056		1,155,956	44,717
Indonesia	1,243,646	+9,687	33,788	+192	1,047,676	162,182
Czechia	1,112,322	+12,486	18,596	+77	987,515	106,211
Netherlands	1,038,156	+3,361	15,017	+88	N/A	N/A
Canada	833,224	+1,647	21,419	+22	778,538	33,267
Portugal	790,885	+2,324	15,649	+127	683,061	92,175
Chile	784,314	+2,275	19,659	+15	743,306	21,349
Romania	768,785	+2,815	19,588	+62	714,709	34,488
Belgium	741,205	+1,717	21,750	+48	50,930	668,525

Israel	737,644	+3,069	5,463	+22	679,133	53,048
Iraq	653,557	+3,575	13,204	+12	611,036	29,317
Sweden	622,102		12,569	+33	N/A	N/A
Pakistan	565,989	+1,165	12,436	+56	528,545	25,008
Philippines	553,424	+1,184	11,577	+53	512,033	29,814
Switzerland	545,535	+1,253	9,838	+21	490,280	45,417
Bangladesh	541,877	+443	8,314	+16	489,254	44,309
Morocco	479,071		8,504		460,628	9,939
Austria	437,874	+1,735	8,290	+30	415,221	14,363
Serbia	426,487	+2,467	4,277	+16	31,536	390,674
Japan	419,015	+1,250	7,102	+87	391,208	20,705
Hungary	391,170	+1,548	13,931	+94	299,989	77,250
Saudi Arabia	373,702	+334	6,445	+4	364,646	2,611
UAE	358,583	+3,452	1,055	+14	343,935	13,593
Jordan	355,106	+2,887	4,503	+12	330,805	19,798
Lebanon	346,080	+2,479	4,152	+60	250,994	90,934
Panama	333,251		5,655		314,797	12,799
Slovakia	282,864	+3,168	6,168	+105	255,300	21,396
Nepal	273,070	+125	2,055		269,394	1,621
Belarus	272,273	+1,352	1,876	+9	261,568	8,829
Malaysia	272,163	+2,998	1,005	+22	229,762	41,396
Ecuador	268,073		15,392		230,377	22,304
Georgia	266,948	+486	3,390	+13	259,727	3,831
Croatia	238,501	+502	5,375	+18	230,907	2,219

By February 17th, 2021, only 19 countries and territories had less than 100 cases out of the more than 200 affected countries and territories. The active cases stand at 22,694,250. Out of these, 22,598,841 (99.6%) are in mild condition while 96,552 (0.4%) are in serious or critical conditions. More than 85 million cases have resulted in recovery/discharge.

MANAGEMENT OF THE CORONAVIRUS DISEASE

There are several strategies in the control of an outbreak. They are: containment, mitigation, and suppression.

Containment is usually done at the early stage of an outbreak. It is done by tracing and isolating those infected with the virus to stop it from spreading to the rest of the population. If the containment of a virus is unachievable, then you would have to move to mitigation.

Mitigation is the measures taken to slow the spread of a virus, as well as mitigate the effect of the virus on the society and health care system. Containment and mitigation measures can, however, be used at the same time in the event of an outbreak.

Suppression is aimed at reducing the basic reproduction number of the virus to less than one to reverse the pandemic. It also requires extreme measures to achieve.

Drastic actions have been taken in an attempt to suppress COVID-19. For example, in China, when the severity of the outbreak became clear, they quarantined cities affecting about 60 million people in Hubei as well as implemented a strict travel ban.

Similarly, South Korea introduced localized quarantines, issuing alerts on the movement of infected people as well as a mass screening of their citizens. Singapore also provided financial support for those who quarantined themselves and imposed a large fine upon those who refused to quarantine themselves. In Taiwan, they increased the production of face masks and placed a penalty on whoever hoarded medical supplies.

Despite the efforts, simulations show that mitigation and suppression have major challenges. Mitigation policies were shown to reduce death by half and peak healthcare demand by two-thirds. However, it would still result in thousands of deaths as well as the health system getting overwhelmed.

Suppression is preferred to mitigation, but it would need to be continued and maintained until a vaccine becomes available. This is because transmissions will quickly rebound when suppression is relaxed. However, long-term suppression will incur social and economic costs.

Managing the Illness

Various COVID-19 vaccines have been given the green light by at least one national body, while many others are still undergoing development and clinical trials. However, no COVID-19 treatment has been developed yet and all medications are currently aimed at treating the symptoms.

When infected with the COVID-19 and you are exhibiting mild symptoms, there are medications to take to help tackle the problem, which are Kleenex®, paracetamol, cough medicine, cold medications, hone, and lemon, Vicks® VapoRub™. Also, having a humidifier will help with flu-like symptoms.

Additionally, if you are asthmatic, it is crucial you have your inhaler, drink a lot of fluids, and rest. You do not have to go to the hospital if you are displaying only mild symptoms, as the hospital is reserved for individuals with severe or critical symptoms.

For serious or critical cases, intravenous fluids, oxygen therapy, and breathing support may be needed. However, the use of steroids may turn out to worsen the outcome. Also, there are still several compounds that may have been approved for the treatment of COVID-19 that are still being investigated.

RESPONSES TO THE CORONAVIRUS DISEASE

SOCIETAL IMPACTS OF CORONAVIRUS

Politics

In China, several administrators in the communist party of China were dismissed over how they handled the quarantine efforts in central China. Also, the Italian government criticized the European Union over their lack of solidarity with coronavirus-affected Italy.

Education

Over a billion children and youth have had their schools affected by the temporary or indefinite school closure mandated by several governments in an attempt to curb the spread of COVID-19. The impact of the closures of schools is more severe for children and families that are at a disadvantage, and also includes childcare problems, interrupted learning, and compromised nutrition. Over the course of 2020, some countries have opened their schools and then closed them again following a surge in COVID-19 cases. As of February 17th, 2021, some countries have opened their schools, but others are still closed.

Socio-economic

The outbreak of COVID-19 has led to several instances of supply shortage, which is caused by the increase in the usage of the equipment to fight the outbreak, disruption in factory operations, and panic buying. In China, the manufacturing hub of the world and a major economy, the outbreak has caused a significant destabilization threat to the economy of the world.

It was estimated that the economic fallout of this pandemic might surpass that of the SARS outbreak, and that ended up being a reality. Also, the outbreak has led to a tremendous decline in oil prices due to low demand from China and other countries primarily infected by the pandemic. Tourism has also been affected by the outbreak due to travel bans and the closure of public places and travel attractions.

Environment

Due to the impact that the pandemic has had on travel and industry, several areas are experiencing a drop in air pollution. Also, the European space agency discovered a decline in the emission of nitrous oxide from power plants, cars, and factories. The pandemic has also caused the waters in canals to clear up, which in turn leads to an increase in fishes and waterfowls.

CHAPTER 4

International Response

TRAVEL RESTRICTIONS

One of the earliest responses to the COVID-19 pandemic was travel restrictions. Initially, countries that had few or no cases spearheaded the travel restrictions by banning entries for people coming from countries that were experiencing the COVID-19 surge. People visiting the country were called upon to quarantine themselves for a period of 14 days as a healthy measure to mitigate the spread of the COVID-19 virus.

However, the virus continued spreading to more countries and regions, leading to the imposition of stricter entry bans and quarantines. In most of the countries that still allowed international travel, people were required to undergo a mandatory 14 days' self-quarantine. In other countries, international travel from certain regions was completely banned.

Besides the travel restrictions imposed by the governments, there was also a decreased willingness to travel as people came to understand the true threat of COVID-19. Particularly, people were reluctant to visit China and some European countries that had been hit the most during the early stages, and later on the United States. Beyond the normal inconveniences that this decreased willingness to travel caused, there were dire social and economic impacts. Air and sea travel companies faced the biggest challenge as inter-country and inter-regional travel threatened to come to a standstill. Countries that depend significantly on tourism faced economic downsides that had not been projected.

Researchers and other stakeholders have debated on the effectiveness of travel restrictions. Most of the specialists agreed that travel restrictions are most useful in two stages: the early stage and the late stage. During the early stage, travel restrictions prevent a virus from being introduced into a new country or region, and this keeps the respective region safe. Similarly, in the late stage, travel restrictions prevent the virus from being reintroduced in a country or region where it has been put under control or no longer existing.

In the COVID-19 case, travel restrictions in the early stage should have been imposed on Wuhan and China to prevent the virus from being spread to new countries. However, the travel restrictions came too late when the virus had already spread to other regions of China and new countries. In the middle stage, when the virus is almost in every country and its rate of infections is surging, travel restrictions are less effective and cannot be applied as the sole technique of containing the spread of the virus, as is the case in the early stage. This occurs because most of the infections are happening internally (in homes, towns, school, work, local gatherings, social circles, etc.) and those emanating from international travel only constitute a small percentage of infections.

EVACUATION OF FOREIGN CITIZENS

In the early stages of the COVID-19 virus, countries were quick to get their citizens from the affected areas. It was still too early to project how the virus would spread, and countries were eager to take away their citizens from the affected countries and bring them home where there we no cases at the time. Besides virus infection, some countries were also concerned that their citizens would face a heightened scope of challenges.

The primary evacuation was for those in the Wuhan and Hubei regions that had been effectively put in lockdown. Prior to evacuation, tests were conducted, and those who tested positive had to be left back as a mitigation measure. Some of the first countries to conduct evacuation included the United States, Canada, India, Australia, Germany, Japan, India, Sri Lanka, New Zealand, Argentina, and Thailand. In most of the evacuations, countries used chartered flights, and clearance was conducted by Chinese authorities.

As the focus of the COVID-19 focused moved away from China to other countries and territories that were experiencing very high infection and death rates, the shift of evacuations also changed. For instance, Canada evacuated 129 Canadian passengers from the Diamond Princess cruise ship after a proliferation of COVID-19 cases aboard the ship. As Iran experienced a surge in COVID-19 cases in early March 2020, the Indian government started evacuating its citizens from Iran. Towards the end of March, the United States followed suit by partially withdrawing its troops from Iraq.

As COVID-19 spread to more countries, the essence of evacuation faded away. With more than 200 countries and territories now experiencing cases of COVID-19, there are no more ongoing evacuations. Countries shifted away from worrying about evacuations to now dealing with the pandemic within the country.

UN AND WHO RESPONSE MEASURES

The United Nations responded through the provision of formal resolutions and operations. The operations were overseen by various UN agencies while formal resolutions were reached upon by both the United Nations Security Council and the General Assembly. The UN Comprehensive Response to COVID-19 was launched in June 2020 by the UN Secretary-General.

WHO is the most significant global body for promoting coordination in health issues, and has played an integral role in helping to mitigate the COVID-19 pandemic. The World Health Organization participated in many areas including raising funds, supply chain, and treatment research. In raising funds, WHO launched the COVID-19 Solidarity Response Fund in March. Numerous companies and several hundred thousand private individuals have contributed to this solidarity fund, including Google and Facebook. By December 7, 2020, the fund had received approximately US$1.5 billion. In addition, WHO created the COVAX vaccine-sharing program to help in the distribution of the COVID-19 vaccine at a reduced or free price.

CHAPTER 5

COVID-19 Mortality

QUANTIFYING MORTALITY

Quantifying mortality is a complex process since various regions use different methods and the numbers vary. For instance, some countries may only categorize COVID-19 deaths based on the patients who die in the hospital. In others, they may categorize the deaths based on those who had previously tested positive for the virus and later on died, irrespective of whether they died in the hospital or at home. Others may decide to only document those who had previously tested positive while others may decide to test even those who have already died without prior testing if they had at all shown similar symptoms. Due to such differences in how the recording is done, the numbers vary from region to region. However, due to the global nature of COVID-19, various measures have been enacted to help inconsistent and standardized quantification of mortality worldwide.

The mortality numbers are also influenced by other aspects, which are different from region to region. These aspects include treatment options, population characteristics, healthcare system quality, the volume of testing, and the time of the first case since the initial outbreak.

MORTALITY RATE

The prevalence of a disease cannot be entirely defined by the number of deaths when evaluating it from a global perspective. Comparing the severity of disease in countries or regions based only on the number of deaths leaves more confusion than understanding. For instance, China has almost 1.4 billion people, the United States has a population of about 328 million, and Israel's population is slightly above 9 million. If each of the three countries has a mortality of 1 million people, categorizing the severity of the disease just based on that quantity without taking into consideration their respective population would be erroneous. To facilitate a better comparison and understanding of the prevalence and severity of a disease, the mortality rate (rather than just mortality) is used.

The mortality rate is calculated by dividing the number of deaths by the population. This can be done for either a specific demographic group, a country, a region, or the entire world. The mortality rate helps in understanding the prevalence and severity of the disease in a given demographic or region. The mortality rate has been the primary method through which the severity of the disease has been evaluated across the world.

The mortality rate is also being commonly used to reflect how the disease is affecting specific demographic groups. This is achieved by dividing the number of deaths in that demographic group by the total population of that demographic group. This has helped to understand the prevalence and severity of COVID-19 in various demographics groups such as the elderly and youths.

INFECTION FATALITY RATE

Not all people who get COVID-19 succumb to the diseases. A great percentage of those who get the virus usually survive it, either naturally or through medication. The statistics of how several infected people survive are very crucial in the research process, as such statistics give hints on the factors leading affecting the death rate. For instance, people who have some underlying conditions such as chronic obstructive pulmonary disease, heart conditions, chronic kidney disease, and cancer are at a higher risk of death after contracting COVID-19. Similarly, elderly people have an increased risk of death compared to young people. To evaluate the number of people dying after being infected, the mortality fatality rate is used.

The infection fatality rate is also referred to as infection fatality risk or infection fatality ratio. It is a common metric that has been used to evaluate the severity of COVID-19 and aid in assessing how it is affecting different demographics. The infection fatality rate is divided calculated by dividing the total number of deaths by the total number of infections.

According to recent systematic review and meta-analysis studies, the infection fatality rate has been different in various countries, and age groups. During the first wave of COVID-19, the infection rate was 0.5–1% in many countries including Netherlands, France, New Zealand, and Portugal. Some countries such as Spain, Australia, Lithuania, and England experienced a 1–2% infection fatality rate. Italy had a higher severity and its infection fatality rate exceeded 2%.

Infection Mortality Rate was lower among younger people and higher among the elderly. The rate for children and younger adults lies between 0.002% and 0.01%. It then increased progressively to 0.4 at age 55. Those at 65 years are experiencing an infection mortality rate of 1.4% which then shoots to 4.6% at age 75 and 15% at age 85.

SEX DIFFERENCES IN COVID-19 MORTALITY

Research studies have acknowledged that some diseases affect people differently based on their sex differences. Some of such diseases that affect women and men differently include Ebola, SARS, HIV, and influenza.

During the early phases of COVID-19, epidemiologic data showed that men and women were being affected differently, with a high mortality rate in men. This epidemiologic data was based on China and Italy. In China, the country's Center for Disease Control and Prevention acknowledged that the death rate for women was 1.7% but men's was higher at 2.8%. While another review confirmed these different mortality rates, later reviews suggested that there was no significant difference in the Case Fatality Rate and susceptibility between males and females.

In an earlier study conducted in Europe, the rate of infection was higher for males at 57% while women's stood at 43%. The mortality was similarly higher among men at 72%, while only 28% for women. Similar scenarios of higher infection and death rates among men have been observed in many countries.

Notably, the sex differences in COVID-19 have not been caused by genetic factors. Rather, they have resulted from lifestyle choices such as smoking and alcohol consumption, which are more prevalent among men than women. Additionally, men develop co-morbid conditions such as hypertension at a younger age compared to women. These underlying factors have been the major contributing factors that have led to higher mortality in men.

ETHNIC DIFFERENCES IN COVID-19 MORTALITY

In the United States, the African Americans ethnicity and other minority groups such as Latinos and Native Americans have received a higher rate of COVID-19 death. However, this high mortality has not resulted from genetic conditions or any biological aspects related to these ethnicities. The main reason behind the high mortality rate among the African Americans and minority groups is structural factors, lack of health insurance, and underlying conditions.

Under the structural factors aspect, most African Americans and other minority groups live in crowded substandard housing, which makes it hard for them to practice social distancing. This had led to a higher rate of infections and proportionally increased mortality. In addition, most of them work in essential occupations such as custodial staff, retail grocery workers, health care workers, and public transit workers. This has predisposed them to COVID-19 infections and led to higher mortality.

Under the health insurance and underlying conditions, most African Americans and other minority groups lack comprehensive health insurance. Most of them are therefore unable to get proper care which has led to an increase in the mortality rate. They also have a greater prevalence of underlying conditions such as heart disease, diabetes, and hypertension. These underlying conditions increase the risk of death among COVID-19 patients and have resulted in a higher mortality rate.

The disparities in COVID-19 mortality based on ethnic differences have also been observed in some other regions including the United Kingdom. An analysis of

COVID-19 mortality in the U.K. shows that blacks, Asians, and other minority groups have experienced a greater proportion of the deaths.

COMORBIDITIES

In the field of medicine, comorbidity refers to the existence of additional conditions in the presence of primary conditions. As such, comorbidities denote the additional conditions that occur on top of the primary conditions. The conditions can either be psychological or physiological and their effect on the primary condition of interest varies.

COVID-19 is a primary condition of interest that has been impacted by many additional conditions (comorbidities). These comorbidities increase the risk for severe illness among the COVID-19 patients. In this case, severe illness for COVID-19 refers to either hospitalization, ICU admission, mechanical ventilation, intubation, or death. Irrespective of age, those with comorbidities have a high risk of severe illness, and most of those who have succumbed to COVID-19 have had pre-existing conditions.

The list below shows the COVID-19 comorbidities that have been established by the Centers for Disease Control and Prevention:

- Cancer
- Down Syndrome
- Pregnancy
- Type 2 diabetes mellitus
- Chronic kidney disease
- Smoking
- Chronic obstructive pulmonary disease
- Sickle cell disease
- Obesity
- Severe obesity
- Heart conditions such as cardiomyopathies and coronary artery disease
- Immunocompromised state from solid organ transplant

Since COVID-19 is still a new disease, there is inadequate data to assess the impact of many other underlying conditions and whether they carry any effect on the risk of severe illness among COVID-19 patients. While the list above comprises of the underlying risks that are now known to increase the risk, there are other conditions that "might" increase the risk but there is no sufficient information to achieve full certainty concerning their direct effects or level of impact. These underlying conditions are:

- Asthma
- Liver disease
- Cystic fibrosis
- Overweight
- Cerebrovascular disease
- Pulmonary fibrosis
- Hypertension/high blood pressure
- Type 1 diabetes mellitus
- Neurological conditions such as dementia
- Thalassemia

These underlying conditions have been the main cause of COVID-19 mortality. In March, data collected in the United States showed that 89% of the COVID-19 patients who had been hospitalized had preexisting conditions. In Italy, 8.8% of the deaths in which medical charts were available shows that 96.1% of those who had succumbed to COVID-19 had at least one comorbidity.

CHAPTER 6

COVID-19 Vaccine

BACKGROUND TO COVID-19 VACCINE DEVELOPMENT

In the history of medical research, no vaccine for infectious disease has ever been produced in less than several years. The feat that has been achieved in developing the COVID-19 vaccine stands at the pinnacle of medical research achievement. However, many underlying factors led to this quick development. The key underlying factor is that there existed many previous projects that had focused on developing vaccines for viruses in the family *coronaviridae.*

Before the COVID-19 pandemic struck the world and facilitated research institutes and pharmaceutical companies in a race to find a vaccine, there had been some ongoing research projects directed at SARS and

MERS. Since COVID-19 belongs to the same group of coronaviruses as SARS and MERS, research institutes and pharmaceutical companies developing the COVID-19 vaccine were at an advantage. They utilized the existing knowledge base about the structure and function of coronaviruses to accelerate the development of the COVID-19 vaccine. All the vaccines have been aimed at providing an acquired immunity against the COVID-19 virus.

The infection rate and ensuing global pandemic resulting from COVID-19 also factored in the quick vaccine development. Due to the urgency of creating a vaccine as infection and deaths continued proliferating, research institutes and pharmaceutical labs had to compress their schedules to reduce the timeline. For instance, some clinical trial steps were combined over months. Under normal circumstances, these clinical trial steps are conducted sequentially over years. Unprecedented collaboration among different research institutes and governments has also expedited vaccine development. In April 2020, the Access to COVID-19 Tools Accelerator initiative was established to help in sharing resources and knowledge about COVID-19 vaccine development.

VACCINE CLINICAL TRIALS

Vaccine research studies started during the early phases of COVID-19. Although faced with many challenges such as rapid development, physical distancing measures, clinical trials, and closing of laboratories, the development has been ground-breaking. By February 2021, only one year and a few months since the onset of COVID-19, there were 66 vaccine candidates in clinical research. The breakdown of the vaccines is as follows:

- 17: Phase I trials
- 23: Phase I–II trials
- 6: Phase II trials
- 20: Phase III trials

Phase I trials encompass testing for the safety and preliminary dosing. Phase II trial evaluates immunogenicity, adverse effects, and dose levels. Phase I–II is concerned with immunogenicity testing and preliminary safety. Phase III comes last and involves vaccine effectiveness and optimal dose.

Phase I involves only a few dozen healthy subjects. Once Phase I is successful, the trial progresses to Phase II, which involves hundreds of people. Phase III, which comes last, involves more participants at multiple sites, including a control group.

TYPES OF VACCINES

Most of the institutes and companies working on developing COVID-19 vaccines have primarily focused on the coronavirus spike protein and its variants. Various options have been explored including non-replicating viral vectors, inactivated virus, recombinant proteins, nucleic acid technologies, live attenuated viruses, and peptides. In all these options, the developing platforms are utilizing "next generation" technologies to enhance the precision and flexibility of the vaccine mechanisms.

By February 2021, 10 vaccines had received authorization for their use from at least one national regulatory authority. The 10 vaccines fall in various categories as outlined below.

RNA Vaccines

- Pfizer-BioNTech vaccine (the United States, Germany)
- Moderna vaccine (the United States)

Conventional Inactivated Vaccines

- BBIBP-CorV from Sinopharm (China)
- BBV152 from Bharat Biotech (India)
- CoronaVac from Sinovac (China)
- WIBP from Sinopharm (China)

Viral Vector Vaccines

- Sputnik V from the Gamaleya Research Institute (Russia)
- Oxford-AstraZeneca vaccine (United Kingdom)
- Ad5-nCoV from CanSino Biologics (China)

Peptide Vaccine

- EpiVacCorona from the Vector Institute (Russia)

EFFECTIVENESS ON THE SARS-COV-2 VARIANTS

As early as August 2020, the first variant of SARS-CoV-2 was confirmed. Since then, a few more variants have been confirmed.

Date of Detection	First Country of Detection	Spread	Classification
Aug 2020	Nigeria	Localized	B.1.1.207
Sep 2020	United Kingdom	Global	B.1.1.7
Oct 2020	Denmark	Likely extinct	-
Dec 2020	South Africa	Global	B.1.351
Jan 2021	Japan, Brazil	Global	P.1

The United States Center for Disease Control and Prevention has claimed that based on the nature of the SARS-CoV-2 virus, it is highly unlikely for a variant to emerge that has the ability to completely escape the immune response. At the moment, most of the vaccines are experiencing reduced effectiveness on the variants, but no variant has been able to completely escape the immune response.

A study conducted by Pfizer, Inc established that there was only a minor reduction of its mRNA vaccine's effectiveness against SARS-CoV-2 variants.

The effectiveness of Oxford-AstraZeneca in the South Africa variant was negatively impacted. Following a study involving 2,000 people, the vaccine showed "disappointing" results against the variant. This led to country to halt the deployment of the Oxford-AstraZeneca vaccine.

Both Moderna and Pfizer vaccines have been found to protect against the UK variant. However, both vaccines have shown less effective in protecting against the South African variant. Moderna is already working to develop a new vaccine aimed at protecting against the South African variant.

DEPLOYMENT OF VACCINES

As of February 15th, 2021, 177.94 million vaccine doses have been administered. This data is based on official reports from national health agencies and has been collated from all countries worldwide. The total number of people who have received at least one dose is 87.75 million. The discrepancy between the number of doses and people who have received the dose is because a vaccination requires two doses.

China has been on the lead in the vaccination process, followed closely by the United States. About 40.5 million people in China have received at least one dose, while the number stands at 37 million in the United States. The United Kingdom comes third at 15 million, followed closely by the EU. Although these countries lead in the number of vaccinations, Gibraltar takes the lead in the percentage of the population that has received at least one dose, standing at 46.3%. It is followed closely by the United Kingdom at 44.2% and Seychelles at 38.6%. The table below shows the vaccine distribution as of February 15th, 2021.

LOCATION	VACCINATED (At least one dose)
United States	37,056,122
China	40,520,000
United Kingdom	15,062,189
EU	13,602,842
India	8,263,858
United Arab Emirates	5,005,264
Brazil	4,946,738
Israel	3,824,480
Russia	3,900,000
Turkey	4,187,763
Italy	3,044,535
Germany	2,635,673

The public has perceived the COVID-19 vaccine differently. Through misinformation and anti-vaccination activists, rumors have been spread globally concerning the impact and side effects of the vaccine. About 10% of the public consider vaccines to be either unnecessary or unsafe, thus leading to the threat of vaccine hesitancy. Research studies aiming to evaluate people's perception of the COVID-19 vaccine have shown that vaccine acceptance has wide disparities by education level, geography, employment status, and race. Vaccine hesitancy is a global health threat since it increases the risk of further viral spread, and thus creates the possibility of new outbreaks.